Stress FreeTM Manufacturing Sol·

I0056720

Ron Mueller

Praise for
Stress Free™ Manufacturing Solutions

*"I read your book and love it. It is wonderful.
The thoroughness and simplicity are excellent.
It deserves to become a best seller."*

Peter Geraets,
European Region Improvement Consultant

Ron Mueller

Stress Free™ Manufacturing Solutions
Root Cause Solutions
for
Manufacturing and Production Systems
By: Ron Mueller

Around the World Publishing LLC
4914 Cooper Road Suite 144
Cincinnati, Ohio 45242-9998

ISBN 13: 978-1-68223-260-6

Distribution by: Ingram
Cover Picture by: Andrey Popov, Dreamstime.com
Cover Design by: Ron Mueller

Ron Mueller

Technical Editor:

Gordon Miller P. E.

Ron Mueller

DEDICATION

To Hien Nguyen Mueller,
the *family problem solver*
and the person
who has enriched my life.

Ron Mueller

Table of Content

Ron Mueller

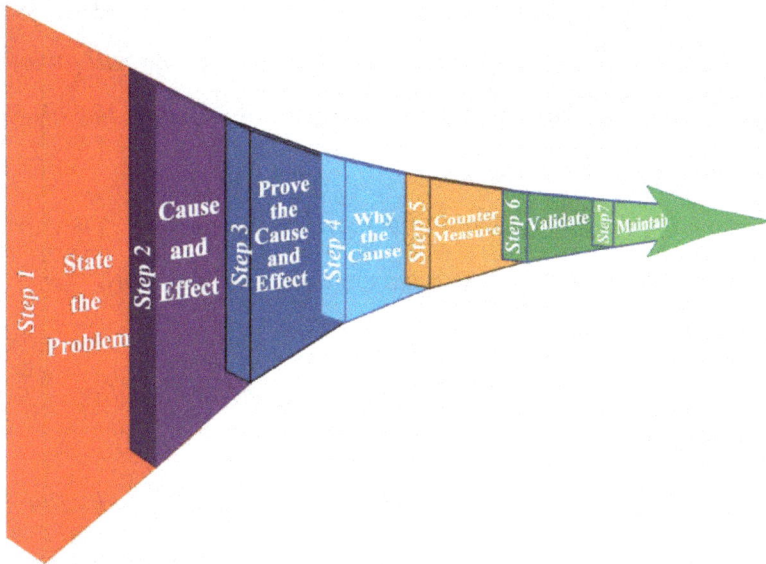

Introduction

You want the problem solved. A world class leader is a world class problem solver who has a series of thoroughly documented solved problems as case studies used to coach others in solving problems.

Do you have your problem solving coaching guide?

There are leaders at every level of the organization. Operating team leaders, Maintenance team leaders, Line Leaders, Area Leaders, Department Leaders, Operations Leaders, Site Leaders, Business Leaders, Lab Leaders; I could go on with the list, but you get the point.

They are all in problem solving roles and they need to be ***great*** problem solvers.

The production line stops. The call for help is made. Demanding customers get less than they ordered. The Vice presidents valiantly give assurances of a quick fix.

Numerous problem solving actions are implemented but the problem reoccurs or in many cases continues un-abetted.

Good intentions but not the required solution.

Then promise of Stress Free™ Manufacturing Solutions, Learning To See with the Mind and the Eye and the promise of "Money to the Bank" is recalled.

"Give me a week and five key production area people. We will give you the root cause solution," replies the **problem solver** *as he brandishes his Stress Free™ Manufacturing Solutions book.*

"I can't free up that many people for that amount of time. They are too busy keeping the line running. Let's get some engineers or those process improvement folks," is a common reply.

"That's fine call me when you all have the people and the time," is the **problem solver's** *reply.*

The people and the time are grudgingly agreed to, and production line cooperation is assured.

The team of five convene. The operator closest to the problem knows it is hopeless. The maintenance technician knows his area will be blamed. The young degreed process engineer figures her career is over. The data hunter resource is willing but confused. And the person responsible for the final resolution has said their prayers and sent out the resumes.

The group is ready to become a team.

The problem persists. The team is asked to define the problem.

The team is asked what is the Phenomenon that must be solved?

"What is a phenomenon," asks one of the team members who gets immediate support from all the rest.

This is team building at its best.

They learn phenomenon and all the rest. The theories are written for each cause and effect. The testing is rigorously done. True causes go into the Why-Why Analysis and countermeasures get identified.

In only five days, the team solves the problem and a few more. Like all heroes they get recognized as the best.

The line keeps running, the product is shipped.

Stress Free™ Manufacturing Improvement

The Stress is over, and people feel free.

Stress Free

Stress Free!   Really?

What makes this problem solving approach stress free?

Solved problems reduce stress.

The approach in this book has been used to solve hundreds of problems all to root cause, reinforcing my statement to you:

"Give me the right five people and any problem can be solved."

Stress Free™ Manufacturing Solutions.

- is a closely guided process that documents each step.
- follows the scientific thought process.
- leverages team knowledge.
- has more than twenty years of a 100% success rate.

"Any problem you can see you can solve."

Stress Free™ Manufacturing Solutions teaches you to **SEE!**

> ***SEE:*** *with your eyes*

> ***SEE:*** *with your mind*

The accompanying:

Stress Free™ Manufacturing Solutions Workbook

Is an excel workbook that guides the user and creates a documented solution for each problem.

Stress Free™ Manufacturing Solutions Examples

Provide solution examples of common manufacturing problems.

The main focus of this book is to teach the user how to focus, how to clarify, how to fully understand a problem and finally how to permanently eliminate a problem.

Problem solving is based on the scientific method.

Stop! Don't run. Stress Free™ Manufacturing Solutions "Keeps it simple".

The scientific method simply implies the laws of physics, of chemistry and of gravity are fundamental.

Any other team will be able to validate the result utilizing the solution test plan. The test will be measurable, thorough, meticulous, precise, and accurate.

Value of the Loss

A key focus for problems solving is the loss recovery expressed in dollars per year. Problem solving may be fun but problems cost businesses significant dollar losses. A key and first step in understanding the problem is to understand the value of the solution. This value should always be front and center for those working on solving the problem.

Defining the value of the loss is critical. The value of the solution needs to be important enough to warrant the effort and cost expenditure of the problem solving team.

Conservatively a problem solving team costs ten thousand dollars per week. The goal is to solve all problems to root cause in one week and to have solutions in place in a month.

The most appropriate way to state the value of a loss is in **dollars lost per year**. Most business cycles and budgets are yearly. The decision to support a problem solving effort must compete with other choices the resources could be working on.

Business Linkage

The dollar loss per year, due to the problem, **must be defined**. It is critically important to create this linkage. What specifically will be gained and how will this gain be measured when the problem is solved.

General improvement goals are certainly important - inventory reduction, speed to market, net outside sales (NOS) gain. However, these are longer term and harder to realize.

Measurable:

- 15 min package change over down to 5 min.
- Work in Process (WIP inventory) reduction from fifteen times the customer "pull rate" (Takt) to two times the customer pull rate.
- Manipulation line staffing reduction from 6 to 2

System Layout

Immediately walking and doing a hand drawing of the system is an important activity that grounds the individual in the understanding of the physical system that they are trying to improve.

It is also important to understand the organizational structure that maintains the current situation. If the improvement change is to be maintained, the organization's leadership must understand how to do so.

The system lay out provides an understanding of.

- People placement and activities
- material placement and handling issues
- current behaviors and practices by individuals
- synchronization issues i.e. - long conveyance, bottlenecks, surge points

It is important for the improvement team to discuss what they have seen and understood from the on the floor walk to make the system layout.

No matter what the problem may be, the leaders of the area and the problem solving team should step back and understand.

- The Customer,
- The Material Flow
- The Equipment
- The Human Connection - how the Work gets done.

Customer Connection

It is critical to get away from the thinking of only meeting the demand of the equipment.

Takt Time is the measure that provides the direct customer pull signal to the production system.

It is worth the struggle to determine how Takt Time is connected to every improvement that is made to the production system. To bypass this struggle makes the improvement temporary.

Takt time should be kept as pure as possible and have no safety/buffer factors associated with it.

Available time:

Calendar or clock time minus planned activities that have blocks of time specifically calling for the production system to be down.

This measure should be kept pure and used to help find the barriers to a pull based production system. Every proposal to change it to accommodate a current measurement approach only serves to hide the improvements required to establish and maintain a flow enabled pull based system.

C/O time relates to inventory reduction. Inventory should only be held to maintain top level customer service. Inventory Reduction is only possible when the system provides reliable flow. Reliable flow is required to support a pull based production system.

Material Connection

The supply of the raw materials and the synchronized handling of the finished product are both critical in ensuring the flow of the production system.

Raw material quantity and placement is critical in optimizing the flow of the production system.

The logistics of bringing the materials to the line and returning material remnants for later use is a critical element the must be analyzed. Flow and time analysis is an important analysis to utilize.

Equipment Connection

Physical layout and material supply layout to the equipment are often related to specific problems or are due to previously experienced problems.

The problem may be the number of stops, or break downs or the amount of effort it takes to keep an area up and running.

There are many opportunities to better utilize the equipment at hand.

- Create flow by the synchronization of filler, capper, labeler case packer by shortening the distances between the equipment to the optimum so the line can start and stop in a synchronous fashion and generate no scrap during a changeover.
- Minimize the raw material inventory in the production area; bottles, caps, KDF's by delivering at Takt or at a pitch.

The equipment layout needs to be drawn to scale. The time that a single product spends having value added and the time it spends traveling needs to be analyzed. The travel time then needs to be eliminated or shortened.

The Human Connection

Once the initial production system drawing is completed then the actions of the people involved in the production of the product should be studied and understood.

Standardized Work.

This is *only the work done directly on the product to produce it to the customer need.* All other work though being done to standard is not considered standardized work. ***Standardized means synchronized to the customer.***

Many organizations produce to a production schedule. This schedule is the equivalent of the "customer". In many cases it is designed to cover the many problems experienced by the logistics associated with moving materials and finished product.

Within this concept people will work to the required standards as defined by safety and product quality requirements and the that required by law.

For continuous improvement to be possible, the critical and repeatable work must be done to standard. The time for this work is normally synchronized with the cycle of the equipment that is being supported. It may also be a service that is synchronized with an accounting or other process cycle.

Stress Free^{TM} Work Solutions is the companion guide in solving problems of manual and office work.

- People and equipment do work.
- Both need to be maintained.
- Both break down.
- Both periodically need fixing.
- People maintain, people fix, people think and solve problems.

Stress Free^{TM} Manufacturing Solutions focuses on the production system, equipment, and material flow.

Realize that given five knowledgeable people you too can solve any problem in the world.

Following the *Stress Free^{TM} Manufacturing Solutions* Process will give you the root cause solution in three to five workdays.

You will always solve your problem.

If not contact: **Ron Mueller**

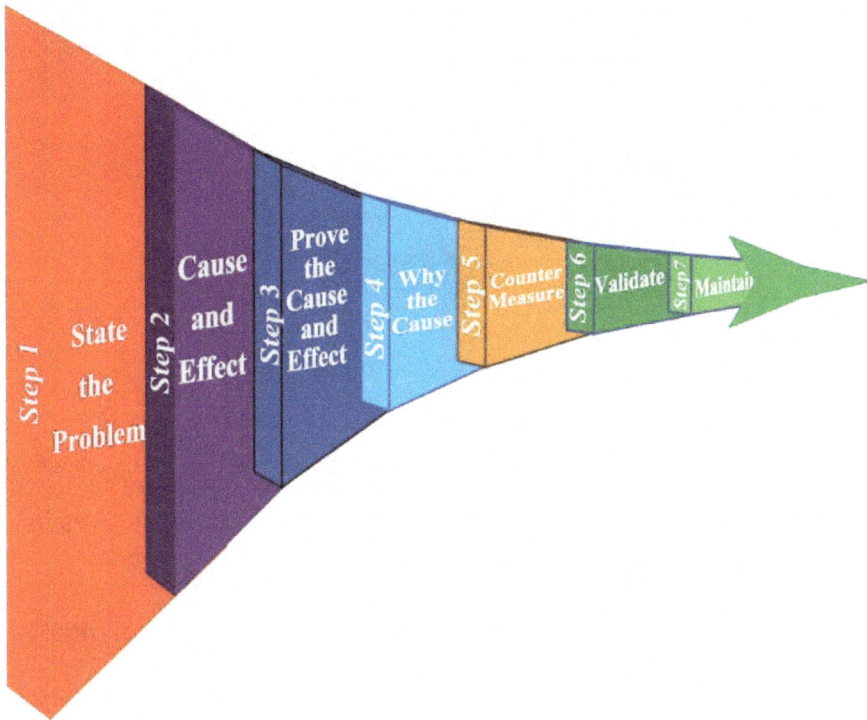

Chapter 1: Problem Understanding, Focus and Solution Value

Before going into problem solving let's talk about; *Measurement*, the *Type of problems*, *True Value* problem selection and how leaders must *Focus* the problem before organizing a problem solving team.

It is important to understand the business cost of the problem before launching into the problem solving process.

A fundamental requirement before beginning any problem solving is the requirement of a *flawless measurement process* or system. In problem solving the ability to exactly measure and have confidence in the measurement is a must. The first problem solving example is one where this fundamental had been overlooked.

A team was called together to solve the problem of missing actives in a product. They worked for almost six months with no progress.

I was asked whether I would add this problem to a problem solving workshop.

"If you give me the right five people," I replied.

I described to them the five people that would be acceptable. There were around ten people involved, including people in China, Japan, and the US.

"The five people must be physically present and only five people," was my reply when the bargaining began.

On the second day of the workshop, I reviewed the cause and effect diagram.

There on one branch was Measurement.

"Measurement can never be a cause," was the point I made.

If you suspect you have a measurement problem, it must be the head of the fishbone or the "effect".

After putting this at the head of the fishbone, the team solved the problem in **two hours**.

Types of problems

Knowing the type of problem allows the leaders of an area know how to select the right people to be on the problems solving team and to determine what tools they would expect a problem solving team to utilize.

There may be other types or more types than are listed but these have served to categorize and organize all the problems I have ever worked on. There are variations and differences in specifics or how a problem may be described but in general they all will end up in one of the six.

1. **Equipment Problems**

   These are problems occurring in the equipment. The equipment function is either reduced or stopped. Action must be taken to re-establish the transformation work the equipment is expected to provide. Often the equipment just needs to be restored to base conditions.

2. **Material Problems**

   These are problems occurring in the material being transformed. The material performance either is inadequate or it prevents the transformation from occurring. The material needs to be evaluated and action taken to ensure the material is within the specification required for easy and consistent transformation.

3. **Processing Problems**

   Processing problems are similar to equipment problems but often they are chemical reaction or flow and mixing oriented. Since visibility is often the problem the analysis tools for these problems aid the investigator by providing visibility of the problem via secondary measures.

4. **System Problems**

   System problems are problems experienced in the management, maintenance, and assurance of the fundamental value stream flow. There are quality requirements, safety requirements, government, and financial requirements. Adherence to these requirements is a fundamental expectation.

5. Human Problems

The problems in this area are problems in how people work together. These problems lend themselves to very different problem solving tools. These types of problems are covered in Around the World Standardized Work.

6. Information Problems

Information flow, timeliness, accuracy, and understandability are critical in holding complex transformation processes together. It is often the single most significant problem in complex transformation processes. There are many ways information can flow.

- Visually – red light to stop cars, signs, symbols, a smile, or frown
- Electronically – internet, e-mail, phone, computer.
- Paper – reports, certificates, orders, receipts, lists
- Audibly – voice, horns, music

Usually, information flows in a mix of the media described. Problems often arise due to incomplete or misinterpreted information.

Problem solving relies heavily on the clarity of information.

All of these problem types have been experienced and solved in every part of the world. They are universal in nature.

True Value Problem Selection

Problem solutions must result in a budget reduction. This is what "Money to the bank" means. The expense of the solution should be quickly recovered.

The problem focusing process breaks what some might think of a single problem worth five million into ten projects worth five hundred thousand each. Key is the understanding that whatever the size of the prize, the team will take the prize to the bank.

It is leadership's responsibility to ensure the problem solving team is solving the most business needed problem (s).

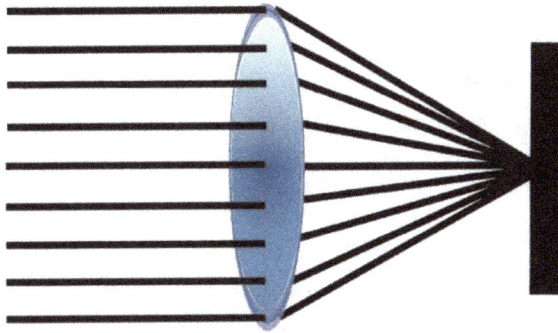

Problem Focus

It is the leader's responsibility to focus a problem or aim the effort before asking a problem solving team to solve a problem. The leader must be a problem solving coach. The leader must have the skill to aim a problem solving team and the ability to ask the questions that help the team define the phenomenon.

The pareto chart is the world's best focusing tool.

The *Pareto Interview Process* (PIP) is simple, fast, requires no data collection, utilizes the knowledge of the "right people", creates organizational alignment and is about 85% accurate.

It is a key tool that helps make problem solving stress free!

Most often I have coupled the *Pareto Interview Process* with the *Material Transformation Analysis* process (MTA).See the chapter: Tools for understanding

When and how to merge the two depends whether the organization has a specific problem on hand or whether they have multiple problems in a production process and are trying to set action priorities.

The ***Pareto Interview Process*** (PIP) for a specific problem is an interview technique that relies on the knowledge of the people in the room.

The facilitator has the simple task of asking two questions.

What is the worst problem you experience?

What is the next worst one and

How does it compare to the worst one that ranks higher?

This initiates the aiming process.

Aiming Process (example)

Step 1: On a chart pad draw the first bar of a pareto.

By definition this problem is the worst and gets a 100% value.

Step 2: Ask: What is the second worst problem?

Ask: Relative to the first how bad is the second problem?

Draw this out on the chart pad.

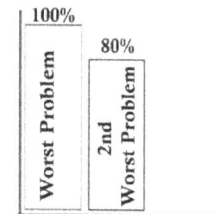

Step 3: Ask: What is next worst problem?

Often the third problem is a significantly lower issue. There may be problems beyond the fourth or the fifth bar but normally they are of significantly lower value.

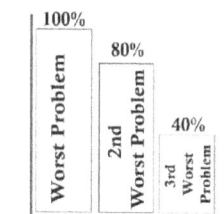

Step 4: Repeat the process for each of the three pareto bars in step three. Go down another two levels if possible.

Step 5: Select the problem at the lowest level that the people in the room choose.

At this point the team often chooses to split up and attack more than one of the pareto bars.

Problem Focus continued

The problem solving team is now *FOCUSED*!

They will walk out knowing two things.

1. Their leader is a capable problem solving coach
2. They have a problem solving process, examples and guidebook that will ensure they solve the defined focused problem.

The number of paretos that are defined may be relatively high.

One problem I helped focus had thirty two paretos for one stated problem.

After the *Aiming Process* I gave the operators the choice of which ONE pareto bar to solve.

The leaders in the area began to argue that all thirty two bars needed solving.

"Solve the one and then come back and talk to me," was my reply as I shut down the discussion.

The reality was that in the process of fixing the one, many of the other bars were eliminated.

At the end of the following half year that area became known for their flawless operation.

Note: The *Pareto Interview Process* (PIP) is often coupled with the *Stress Free^{TM} Material Transformation Process* to identify the worst ingredient, then the worst transformation, then the worst work point in a production process. This is a way to use the hands on knowledge of those at the point of cause to clarify and aim improvement efforts.

See Chapter 4: Problem Solving tools.

Ron Mueller

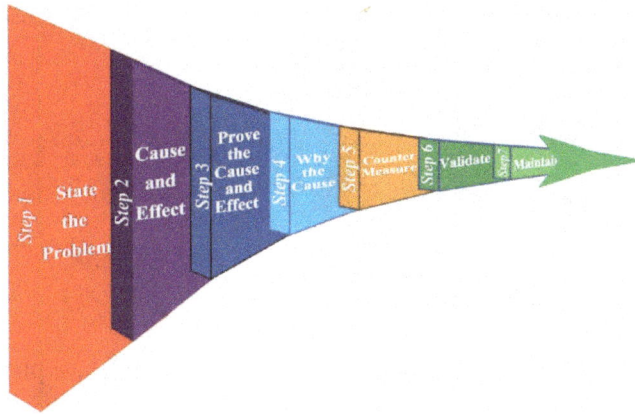

Chapter 2: The Problem Solving Process

Most problem solving processes are similar. They vary in the number of steps and what the steps are called but in the end they seek to identify a root cause for a specific problem.

Stress Free Problem Solving reduces all the approaches to a single, simple, and very successful one.

Pieces, parts, and practices from all processes are acceptable. I never argue with someone wanting to use a tool they know how to use. In the end all have put those tools away and have utilized the simple approach presented.

The Stress Free Problem Solving Process.

Step 1: State the Problem.
Step 2: Cause and Effect
Step 3: Prove the Cause and Effect
Step 4: Why the Cause
Step 5: Countermeasure
Step 6: Validate
Step 7: Maintain

Stress Free Problem Solving

This process is available in an Excel workbook.

The workbook provides a way to organize the problem solving effort.

It is an aid to standardize the approach and provides a way to maintain discipline essential in problem solving.

See the example in the back of this book.

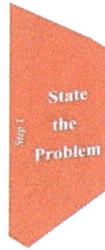

Step 1: State the Problem

To understand the situation the problem statement must be precise. The problem statement must focus on only one *phenomenon*. This is a big word with a simple meaning.

Phenomenon: **What you see: Singular**

This simple concept seems to be the hardest for problem solvers to accept. The problem examples in this book were solved when problem solvers were able to focus on a single phenomenon.

So, the problem is worded to *what is seen*.

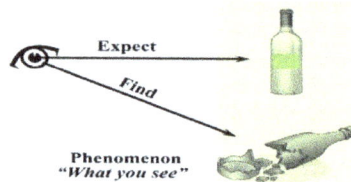

The statement of the problem often gets mixed with other important but distracting information. The observer of the problem must state only what is seen. To say anything else projects solution thinking into the understanding of what the problem's root cause might be.

Knowledge and understanding about what is seen must be developed.

The situation must be investigated.

Stress Free^{TM} Manufacturing Improvement

There are a several important knowledge development tools.

1. **Material Transformation Analysis** (MTA)[1]
    This tool examines the transformation of each material that is in the product. The transformation point where the problem exists can then be further examined using work point analysis.

2. **Work Point Analysis** (WPA)[2]
    Work Point Analysis examines the elements that are associated with each work point.

3. **Direct Observation Sheet and Stopwatch** (OSS)[3]
    This is used to understand the human to equipment and material interaction.

4. **Travel Chart**[4]
    This is used to understand the human movement around the work area.

These tools must be practiced out on the floor. They are useful only to the problem solver if they get used "on the floor" during the problem solving effort.

The saying is,

"If you haven't seen it. You are not allowed to talk about it." [5]

When the phenomenon statement can be stated with simple, *what you see* clarity, the problem will be solved to root cause.

These tools are in the Stress Free Problem solving workbook and shown later in the examples in this book.

*1 This is an industrial engineering tool modified to material transformation system evaluation

*2 WPA from TPM and is invaluable in determining the 100% quality condition for each component in a production system.

*3 OSS is used to observe the individual

*4 Follow someone, draw a line for each from here to there movement. See what you get.

*5 From Toyota stories by Satoko-san.

Step 2: Tips on Stating the Problem

1. Keep asking those involved in the problem,
2. What do you SEE?
3. If the problem is inside where one can't see the problem then ask, "What is your theory or problem model?
4. If there are competing theories, accept them and have each person postulating the theory write down their thinking.
5. Make sure the learning tools get thoroughly exploited and that they get use "out on the floor".
6. Utilize use photographs and video to better see the problem.

Step 2: Tools for Aim and Understanding

1. Material Transformation Analysis
2. Pareto Analysis

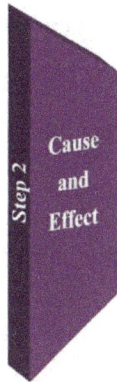

Step 2: Cause and Effect (C&E)

This is the next most critical step.

- The Effect, the problem or phenomenon defined in step 1 is the focus.
- Each cause MUST have a **direct** relationship to the Effect.

Effect: The bottle is broken.

Cause:

  - the bottle was dropped
  - a rock hit the bottle
  - the bullet from the gun hit the bottle

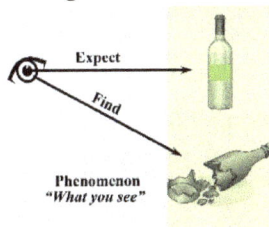

The cause and effect relationship is always written as if true.

Cause and Effect Workbook Cause input lists.

These lists automatically fill in the cause and effect worksheets associated with each list item.

Most input into the worksheets will only need to be done once. It will then be copied into all forms that use the same information.

Steps 2.1 through 2.5 are the input areas below.

This is about identifying the CAUSE. The cause must have a direct impact in creating the EFFECT or PROBLEM.

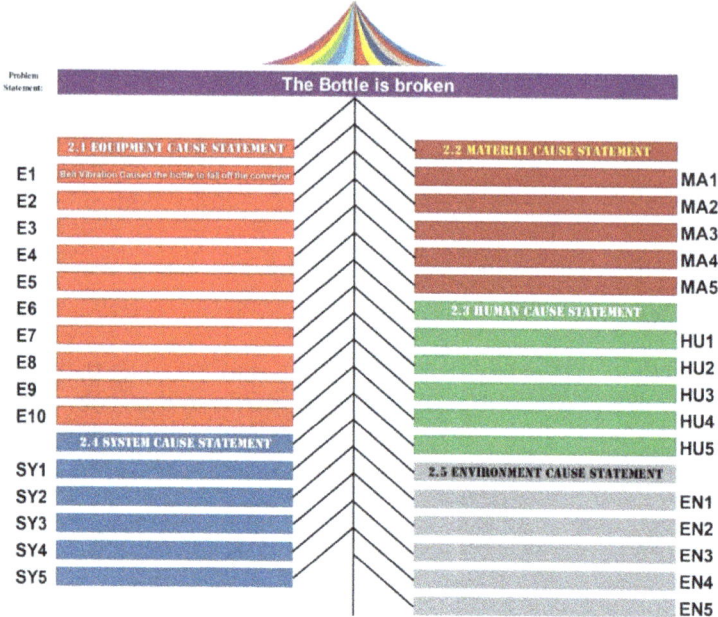

Step 2.6 Create Cause and Effect proof of Cause worksheets.

Equip 1	Material 1	System 1
Equip 2	Material 2	System 2
Equip 3	Material 3	System 3
Equip 4	Material 4	System 4
Equip 5	Material 5	System 5
Equip 6	Human 1	Environment 1
Equip 7	Human 2	Environment 2
Equip 8	Human 3	Environment 3
Equip 9	Human 4	Environment 4
Equip 10	Human 5	Environment 5

This is the central PRINT SELECTION AREA

Step 2: Cause & Effect (C&E) (continued)

This Step provides a link to each cause and effect worksheet that must be filled out.

Cause and Effect Worksheet	
Cause statement	This comes from each input area. (Equipment, Material, Human, System and Environment
Effect Description	This is what you see - the broken bottle.
Standards	Exist Name Applied
	If standards exist, then a check of adherence to standards is the first step. If no standard exists, then at the end of the problem solving a standarrd is often set for that solution
Cause Thesis / Theory	The Cause Thesis or theory is written in the positive sense. If proven true then this cause has a contribution to the effect. If the theory proves to be false this particular cause is put aside.
Thesis Diagram	A visual diagram often aids in explaining the thesis.
Proof / Test Plan	This is step by step test plan. The intent is that it can be validated by anyone following the steps. This allows for follow up if there is any disagreement.
Data	Fresh, Obtained from the source
	From Computer System
	From the Expert
Action	Immediate corrective action is the best practice. This allows for quick verification of the solution.

Example Cause and Effect Worksheet

This form is used over and over for each type of Cause. It documents the work done and should be kept as the record or the case study documentation for each problem

Step 2: Tips on Stating the Cause and Effect

1. Test each cause and effect relationship.
   "Because of this – the problem happened.
2. Write each cause and effect relationship. Then ask the problem solving team members to analyze the statements for clarity.
3. If there are competing or opposite theories, accept them and have each person postulating the theory write down their thinking.
   > Do NOT debate the opposing theories. Accept both and Allow the individuals the opportunity to learn.
4. Whenever possible draw the cause and effect relationship.
5. Use engineering documents or pictures to describe the problem.

Step 2: Tools for Cause and Effect

1. Field observation
2. Travel Chart
3. Scientific thought process

Step 3: Prove the Cause and Effect

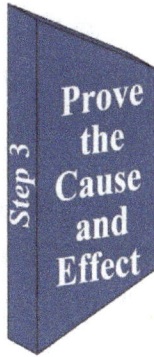

"Because of this cause, the observed effect happens."

The theory of how the cause creates the effect is written down in the work sheet shown. This is the thesis or theory that describes the relationship between cause and effect.

Often two people have opposing theories. This process facilitates each writing their theory and then designing a test to prove it to be true.

A thesis diagram provides a visual, showing the relationship between cause and effect. This visual often is all that is needed to align the problem solvers on the problem.

The proof or test plan is the way defined to verify the truth of the thesis.

The importance of being rigorous in this proof cannot be overstated.

- Each test plan must be measurable, verifiable, and repeatable.
- Done well this step greatly focuses the problem down to a few or only one cause.
- Every Cause and Effect work sheet must be rigorously field tested.

Step 3: Tips on Proving the Cause and Effect

1. Field verify the test plan by a simulated walk through.
2. Adjust the test plan based on field learning and operational discussions.
3. Make arrangements with the on the floor operation for their participation.
4. Ensure all necessary data will be gathered and preserved. If product or material samples are to be analyzed make sure a sampling plan exists and includes marking each sample.
5. Do a thorough results evaluation. Whenever possible make the evaluation a "statistically valid" one.

Step 3: Tools for Proving Cause and Effect

1. Test Plan
2. Field Observation
3. Control Charting
4. Statistical Analysis

Step 4: Why the Cause

Theories proven to be **true** in the Cause and Effect process are brought into the Why–Why analysis process.

- The Cause is the item in the left hand column of the Why-Why worksheet.

The question then is why does this cause occur?

- There may be several first level whys. Each first level why must all be field checked and verified **before** the second level whys are asked.
- The Why-Why is executed in a vertical direction

The process of asking the next level of why continues as long as the next level can be clearly stated.

- In most cases when the Cause and Effect has been rigorously executed, it is very difficult to go beyond the third level Why.
- Each first Why Level must be field checked before defining the second level whys. The same approach is then followed for each subsequent why level.

Getting out to the fifth why is often a sign of a not properly stated effect or phenomenon.

The last two columns identify who is responsible for executing the root cause correction and by when it will be done.

If you get beyond three why's with any of the true cause and effect worksheets that were true, it is a sign that you did not do as thorough and rigorous cause and effect analysis as you should have. The benefit of the why-why at this point is that it makes your mind look at the cause in an almost opposite direction. This causes one to see some additional reasons of why the problem existed.

Additionally, it captures all the counter measures in one location and creates a countermeasure action plan with dates and people responsible for the countermeasure.

The right hand side of this chart is the root cause

Step 4: Tips on Why-Why and Root Cause

1. Make sure each true Cause and Effect is taken through Why-Why. Do this even if "Root Cause" has been found during the use of Cause and Effect.
2. Why-Why is the mirror image of Cause and Effect. It causes one's mind to think about the problem differently.
3. Each Root cause when implemented should eliminate the final why in the why-why analysis.
4. Training all personnel is always a required action. It is not a root cause.
    Remember "Direct Cause and Effect Relationship,"
5. The form shown has steps 5 and 6 included. Often this is all the "plan" that is needed.

Step 4: Tools for Why-Why

1. Field Observation
2. Control Charting
3. Statistical Analysis

Step 5: Countermeasure

It is critical to execute the countermeasure and then validate the elimination of the problem.

Many times, the Why-Why form is sufficient as a countermeasure plan.

On larger or longer solution execution time frames a counter measure plan with measurable milestones is critical.

Safety should be verified before counter measure implementation.

Countermeasure Plan Example format

- the specific fix to physical objects or chemical reactions.
- that standards will be written
- that the storeroom will be updated based on the countermeasure
- maintenance and operating procedures will be appropriately adjusted.
- training of everyone associated with the counter measure will occur.
- validation and follow up will occur

Step 5: Tips on Counter Measure Implementation

1. Make an immediate improvement.
2. Act immediately, standardize or meet system requirements over time.
3. Make sure all shifts know about the countermeasure.
4. Train all personnel on the countermeasure.
5. Follow up on the countermeasure implementation.

Step 5: Tools for Counter Measure Implementation

1. Project Plan
2. Maintenance Calendar and Planning System
3. Daily Log

Step 6: Validate

The "goodness and ease of maintaining" a root cause solution directly affects the probability that the root cause loss free condition will be maintained.

	Questions	Evaluation					
	Equipment	No	Not Sure	OK	Mostly	Good	Best in Class
1	Equipment operating condition is easy to see.						a
2	Equipment operational and physical settings are easy to set.						a
3	Equipment settings are fixed for each specific product.						a
4	Production variations are easy to prevent.						a
5	Operational centerlines are automatically maintained.						a
	Material	No	Not Sure	OK	Mostly	Good	Best in Class
6	Material Quality Characteristics have units and tolerances defined.					a	
7	The supplier is certified and the material Cpk>1.33					a	
8	Only logistical damage inspection is required of incoming materials.					a	
9	Damaged incoming materials are not unloaded.					a	
10	In process material issues are rare and easy to recover from.					a	
	Human	No	Not Sure	OK	Mostly	Good	Best in Class
11	Expert problem prevention skills are the norm.						a
12	Advanced condition management skills are the norm.						a
13	Adherence to Standards and following Standard Operating Procedures is the norm.						a
14	Mistake proofing has eliminated mistakes.						a
15	Our people are highly motivated, continuous improvement leaders						a
	Systems	No	Not Sure	OK	Mostly	Good	Best in Class
16	Systems exist and are well documented.						a
17	Systems ensure Cpk>1.33 and no defects						a
18	Systems ensure required production rate is maintained.						a
19	Systems are explained and trained using the lastest techniques						a
20	Systems are easy to learn requiring less than five days to get qualified.						a
	Environment	No	Not Sure	OK	Mostly	Good	Best in Class
21	Support Organizations accept our solutions and integrate it into their design.						a
22	Purchasing adjusts its buying based on the problem solution.						a
23	Material specifications are adjusted based on specific problem solutions.						a
24	Operational conditions are changed as required to maintain improvements						a
25	Logitics are adjusted to support specific problem resolution.						a
	Leadership	No	Not Sure	OK	Mostly	Good	Best in Class
26	Leaders review the root cause of all key problems.						a
27	Key problem resolution always get public recognition.						a
28	Leaders schedule root cause validation time.						a
29	Leadership goes to the root cause to see and get understanding.						a
30	Leaders participate in stress free problem evaluation.						a

Statistical performance or correction validation is one of the best ways to ensure the solution is the root cause.

Visual Stress Free Evaluation

Stress Free Solution Score
Minimum Goal is 80 %

- Best in Class Solution
- Maintaining Solution requires little effort
- Solution is System Supported, some effort to Maintain
- Periodic Checking is required to maintain solution
- Solution is hard to maintain
- Stress Free Solution Rating

80.0%

I have had what I believed were phenomenal solutions that had low evaluations. They required the constant attention of the operators. They do not appreciate the solution for very long if they must attend to it on a daily basis.

These types of Solutions are acceptable but should then be put on checklist for future improvement.

Step 6: Validation Tips

1. Enroll affected personnel in the evaluation.
2. Listen to what the conversation during the evaluation.
3. Take notes of suggested improvements.
4. Use Cause and Effect to capture suggested improvements.
5. Have the people involved try out the improvements.
6. Set a fixed time to call an end, celebrate the success and focus on the next critical problem.

Step 6: Tools for Validation

1. Statistical Analysis
2. Control Chart
3. Time

Step 7: Maintain

All documents and drawings affected by the countermeasure must be updated. All operating procedures must be modified to reflect the counter measure.

Each solution should have a standard that maintains the solution.

When everything is in order everyone associated with the problem and its solution must be trained and if necessary qualified in the new procedure.

Maintain Items

- Scheduled training and qualification
- Periodic Re-retraining and re-qualification
- Maintenance Calendar
- Maintenance Contract updates
- Operational Calendar
- Periodic Inspection
- Engineering records update
- Purchasing/Spare parts records update
- Visualization

Step 7: Tips on Maintain

1. Use the hand edited field paperwork immediately. The formal paperwork may take longer to get through the system. Assign an owner to update all documents.
2. Make sure the spare parts area gets the information on any parts that have changed and if anything was made obsolete.
3. Make sure all training documents get changed appropriately.
4. Make sure all maintenance procedures and documents are addressed.
5. Schedule the requalification of all people as required.
6. Train, train, train

Step 7: Tools for Maintaining

1. Maintenance Calendar
2. Training
3. Periodic requalification of people and equipment

Ron Mueller

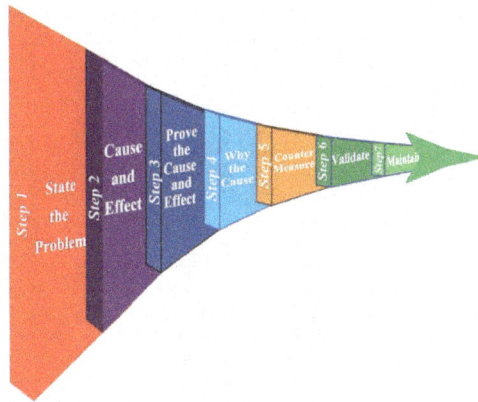

Chapter 3: Solved Example from Workbook

The makers of Great Chips experienced a quality problem with their packaging. Consumers complained that the artwork was smudging off. This was a problem that needed immediate attention.

A problem solving team was put together to solve the problem. The team coach was an expert problem solver. He guided the team to the use of Stress Free™ Manufacturing Solutions book, and its accompanying Excel Workbook. Guide.

The first step was to clarify what the problem or phenomenon that needed to be solved. The team studied the defects that had been obtained. They then went on to evaluate the process of making the package and the printing of the artwork on the package.

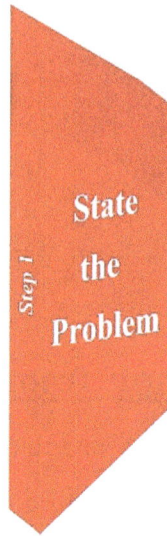

Step 1
State
the
Problem

The team flow charted the packaging material. The three major steps were Extrusion Blowing, Printing and Finishing. Each of these major steps had sub processes that might contribute to the phenomenon.

The team followed the PPS guide and spent some time clarifying the Phenomenon.

Step 1: State the Problem

Step 1: State the Problem

A better description was the ink was not bonding to the polyethylene plastic.

The next step was to postulate the potential causes of the phenomenon (problem).

The team was guided to identify the potentially direct causes of the ink not bonding properly to the polyethylene plastic.

Their problem solving coach made several key points about the cause and effect exercise.

1. The relationship between cause and effect had to be **DIRECT**.
2. Since the problem was physical neither **M**ethod nor hu**M**an could be a direct cause.
3. The environment might be a contributing factor but again it was not a direct cause.

The direct relationship requirement greatly simplifies the identification of causes. This is not brainstorming for new ideas. The intent is to clearly identify all potential causes that can directly cause the problem.

There may be contributing factors that create the situation that result in the cause, but these will be addressed later as part of the solution as counter measures to prevent the recurrence of the problem.

It is a scientific cause and effect. These worksheets each test the cause and effect hypothesis using scientific principles.

E1	equipMent Cause Proof Worksheet
Cause statement	Corona power goes low
Effect Description	Ink Printing is not bonding to the polyethylene plastic
Cause Thesis / Theory	Corona treating increases the surface energy of plastic film, to improve wettability and adhesion of inks A dip in the corona power will lead to spots of untreated poly where ink will not properly bond
Thesis Diagram	
Proof / Test Plan	1. Check history of the Corona System 2. Induce the situation most suspect 3. Evaluate results

	Exist	Name		Applied
Standards	Yes	Corona System Centerline CBA		Yes

Data

Fresh, Optained from the source
From Computer System
From the Expert

Enter True Or False > TRUE

Action
Corona Low Power caused the film treatment to be negatively affected.

- The cause and effect worksheet provides a focused document that brings all the information together on one sheet.
- The Cause and Effect sheet becomes the record for later review and for ensuring the right work is being done to prevent recurrence of the same problem.

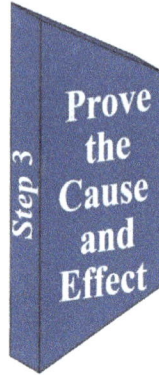

There were seven equipment, three material and two environmental cause thesis postulated. Each was clarified using a cause and effect worksheet. The worksheet begins with the cause statement. It is additionally clarified with the effect statement.

The next block is where the scientific thought process is put into action. The person proposing the cause must write up his or her thesis or theory of how the cause can create the effect.

Often there are competing theories. This is OK. Have each person write up their theory. They will then test their theory. They will either prove the theory true or they will prove it false.

The act of drawing out the thesis diagram and the accompanying theory test plan often resolves competing ideas. If not, the subsequent scientific test will do the trick.

Before going too far into the problem solving, check whether standards that affect the problem exist and that the standards are being followed. If they do exist and are not being followed, the first step is to re-establish the standards and check to see if the problem is solved.

In this case the history of the corona system power was checked. The loss of power and some periods of low power were documented in the historic production system data. This seemed to support the theory.

The coach suggested that low voltage should be induced on the corona system. This low voltage test created the situation where the printing ink did not properly bond to the polyethylene surface. The theory had been proven to be true.

False theories die immediately.
True theories go on into why-why analysis.

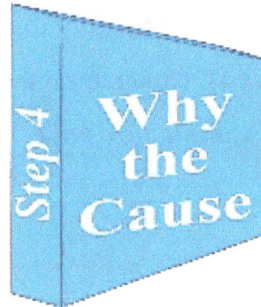

Step 4 Why the Cause

The problem solving coach pointed out the reverse relationship of cause and effect and why-why. This reverse relationship often causes the problem solver to see the situation in a different light.

The true cause is tested as to why it happens. In each case the test is a fresh in the field test. These tests should be managed to minimize the loss, but it is critical to learn if the problem can be turned on and off.

When this ability to turn the problem on and off is achieved, the countermeasures are very precise and effective.

Step 4. WHY-WHY ANALYSIS							Action	
True Cause & Effect	Why 1		Why 2		Why 3	Root Cause	Counter Measure	Date	Resp.
Equipment									
Corona power goes low	Low Corona Power	True	Power System low power	True		Power System low power	Corona System Power Back up- Investigate cost Reject product when power system dips	Immediate	Jose Maria
			Power System interuption	True		Power System interuption	Reject product when power system interuption cost	Immediate	Maria
			Loose power connection						

The action may be as in this case one that does not resolve the root cause but makes practical sense for the business. This does sound a bit off the root cause mantra but let's be clear. This is a business oriented problem solving process and not about the purity of the process.

Having no power dips requires the investment in an un-interruptible power supply. In this situation the power company was not dependable and having no power dips requires the investment in an un-interruptible power supply. This requires investigation into the cost benefit of the power supply.

A red light and an automatic poly position logging system was implemented to indicate a low power condition and to reject the poly made during that period.

This solution greatly improved the ability to deliver high quality poly and not make the immediate high cost expenditure the company could not afford.

This is the End of this Example. Step 5,6,7 are not included.

Chapter 4: Problem Solving Tools

This chapter presents an ***array of tools*** that are used to develop focus understanding and help guide the problem solver to root cause.

Not all the tools have been used in the examples in this book but have been employed in solving the many problems addressed by the process and the reader should be aware of them and pull them in as necessary when needed.

Tool For Focusing

Pareto Analysis:

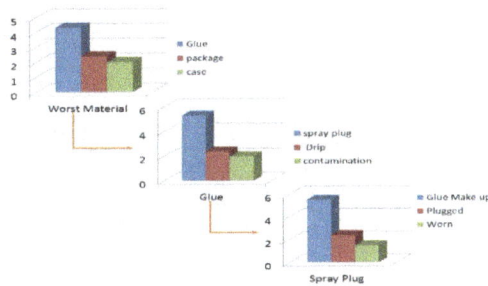

The Key step in clarifying the problem is to quickly establish a cascading Pareto.

In this example ask, "What is the worst material?"

The agreed to answer becomes the first Pareto bar.

Then continue to ask about the second, third materials. Always proportion the next worst to the one before it.

Repeat this process to identify the worst problem within the worst material.

Do this to create the three sets of paretos.

Notice how it aims the process.

Tools For understanding
Material Transformation Analysis:

Material Transformation analysis is one of the most powerful ways to quickly learn how a product is produced. Each ingredient in the product is identified and placed down the left side of the paper. Then the transformation of each ingredient is shown. How each ingredient is combined to make the product is represented by conveyance lines and arrows, transformation points, inspection points, delay points etc.

The entire diagram provides a means of seeing very complex processes in a very simple way.

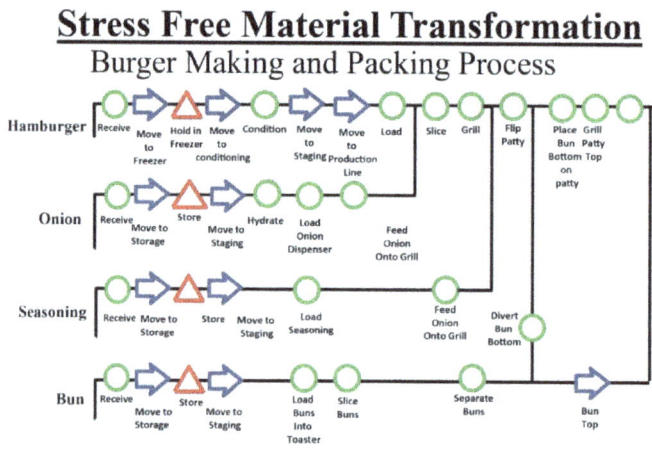

Stress Free Material Transformation
Burger Making and Packing Process

Only the green elements add value.

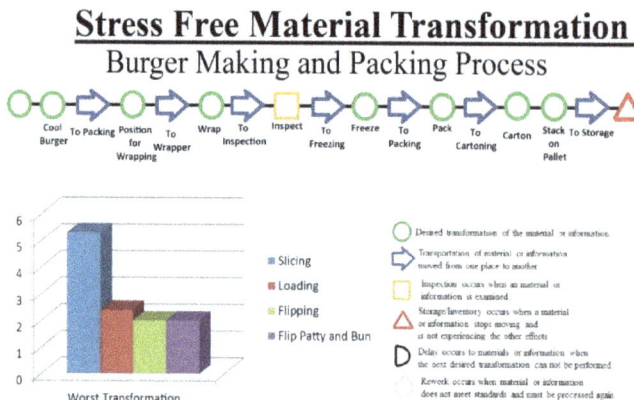

Stress Free Material Transformation
Burger Making and Packing Process

Tools For understanding

Value Stream Mapping:

This tool provides essential information of the status of the value stream. Key elements on a value stream map are:

Material flow, quantity, and time.

Information flow and to whom.

Transformation points

Number of people in the process and where

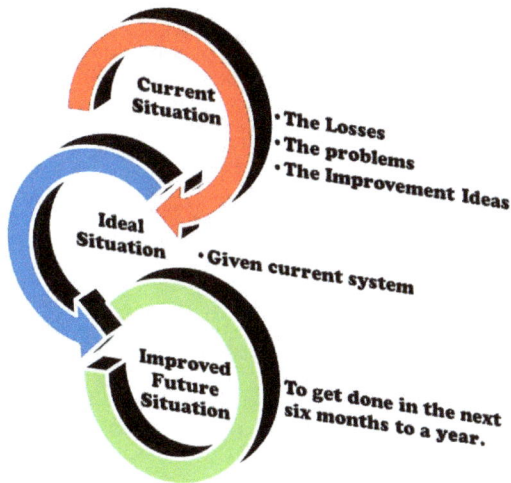

Time Flow Analysis:

The focus and the visual associated with this analysis is material movement and transformation time in seconds or minutes. The amount of time material spends in the system at each point along its travel indicates where delays and inefficiencies exist.

Current Time Flow Analysis — 240 Supply Chain Days

Transport
Inspect
QA Inspect
Wait for Truck

Supplier Inventory 45 days
Site Raw and Packing Materials Inventory 30 days
Distribution Hub Inventory 85 days
Customer Distribution Inventory 50 days
Store Inventory 30 days

Future State Time Flow Design — 140 Supply Chain Days

Supplier Inventory 30 days
Site Raw and Packing Materials Inventory 15 days
Distribution Hub Inventory 40 days
Customer Distribution Inventory 30days
Store Inventory 25 days

Legend:

- Desired transformation of the material or information
- Transportation of material or information moved from one place to another
- Inspection occurs when an material or information is examined
- Storage Inventory occurs when a material or information stops moving and is not experiencing the other effects
- Delay occurs to materials or information when the next desired transformation can not be performed
- Rework occurs when material or information does not meet standards and must be processed again

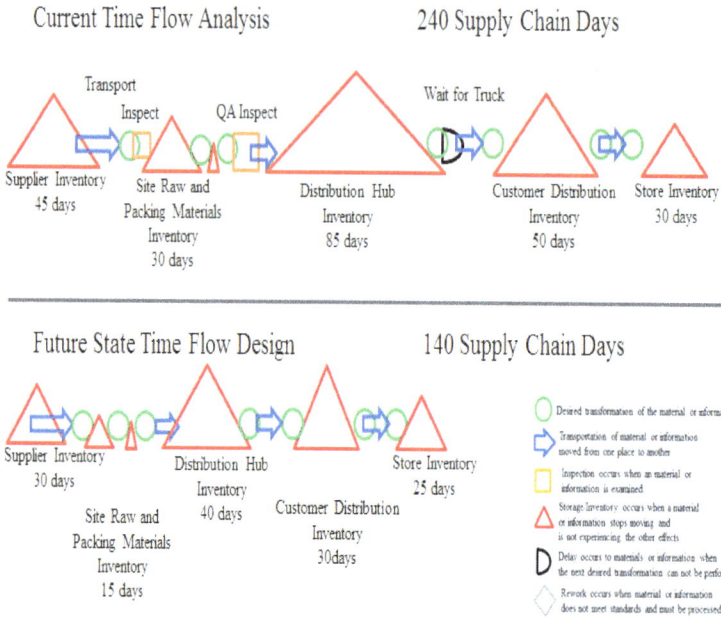

Tools For understanding

Work Point Analysis:

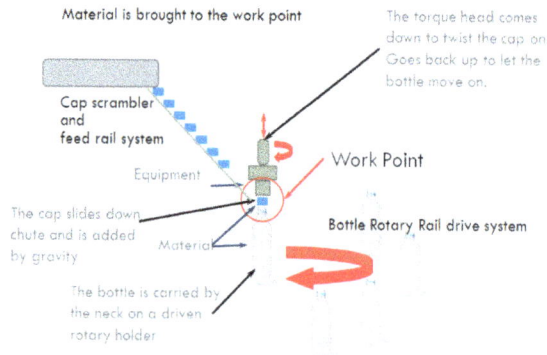

Material is brought to the work point

The torque head comes down to twist the cap on. Goes back up to let the bottle move on.

Cap scrambler and feed rail system

Equipment

Work Point

The cap slides down chute and is added by gravity

Material

Bottle Rotary Rail drive system

The bottle is carried by the neck on a driven rotary holder

This analysis produces a zero loss component inspection check list. This check list covers all systems and components associated with the material transformation into the product desired.

Keeping all components within tolerance will result in zero downtime or quality problems.

Tools For understanding
Work Point Analysis:

This form guides its user to create a new deeper understanding for the transformation work point.

Work Point Summary Form

1. **Work Point:** *Where the equipment and materials make contact to transform "Work"*

2. **Operating Principle:** *What happens as the product and equipment make contact at this point. How does it work?*

3. **Standards:** *List known standards for this point; CIL, centerline checks, etc.*

4. **Loss Impact:** *Loss that occurs at this work point*

5. **Define Work Point:** *Where the product & equipment make contact – show main structure and components*

6. **Sketch the Work Point** *Draw or show point of contact diagram*

   Equipment Components:

   Product/Material Components:

   Process Conditions / Variables:
   Temperature, pressure, Flow, Speed etc.

   Processing Aides
   Air-blow, anti-friction spray, catalysts, etc.

Step 7: Determining Work Point POSITION	
Fasteners	
Piping	
Rotating	
Lubrication	
Pneumatics/ Hydraulics	
Electrical & Instruments	
Static Elements	

Step 8: Determining Work Point CONTINUITY	
Fasteners	
Piping	
Rotating	
Lubrication	
Pneumatics/ Hydraulics	
Electrical & Instruments	
Static Elements	

Tools For Root Cause Identification

Cause and Effect Analysis:

Cause and Effect Analysis provides a simple means for identifying multiple contributors to a specific phenomenon. It is one of the primary tools utilized in the examples shown in this book.

It is especially useful on complex problems since it allows the elements of the problem to be examined one at a time. This provides the people on the floor a simple means of addressing very complex problems.

By using the problem solving approach presented in this book, multivariable problems requiring complex design of experiments is reduced to minimum.

Tools For Root Cause Identification

Why-Why Analysis

Why-Why analysis was developed in the 1940's in the US. Procter and Gamble has a 1947 copyright with this process described.

It came back to the US in the late 1980's and 1990's from Japan and the Quality Circle focus.

It can be used by itself. However, my experience with Why-Why analysis causes me to recommend the use of Cause and Effect first followed by Why-Why analysis. This approach ensures a focused Why-Why. It reduces Why-Why analysis often shown on a C size drawing to a normal 8.5 by 11 inch paper size.

This is an example from the Stress Free Problem Solving Workbook.

For Root Cause Identification

Phenomenon Mechanism (PM) Analysis

Phenomenon Mechanism Analysis engages simple statistics and chemistry to analyze a problem at a more detailed level. It is similar to moving a microscope slide from a 10 x magnification to a 100 x magnification. The problem or phenomenon definition does not change. However, the description of the problem must now be described in the terms of physics, force diagrams and equations.

PM analysis is required in less than 2% of the problems if the Stress Free Problem solving process is rigorously followed.

Example:

Goal: serve a full cup of tea with no spillage.

Phenomena: 7% of time some tea is spilled outside the cup

Phenomenon 1   Phenomenon 2

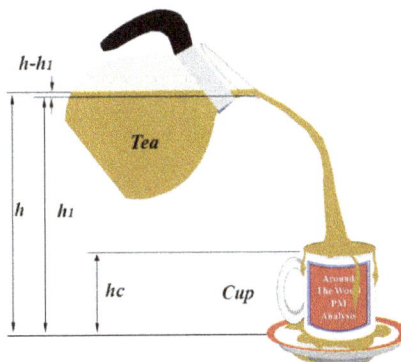

There are two phenomena that result in tea outside of the cup. The analysis of these phenomena using the basic laws of physics allows for the exact definition of all the contributing factors leading to tea outside of the cup.

Ron Mueller

Chapter 5: Stress Free™ Problem Solving Workbook

The companion *Excel Stress Free Manufacturing Problem Solving Workbook* provides a guided way to successfully analyze, solve and thoroughly document a problem.

Around the World Publishing Website; www.ATWP.US

There the excel workbook can be purchased via PayPal and immediately downloaded.

A series of Solved problem workbooks are also available. These solved problems are ones that may be similar to a problem you are currently facing. The intent is to provide a solution that may accelerate the problem solver in attaining their solution.

Entry page for the Excel Spread Sheet

Each step is linked to the work area for that step. This is a fully linked workbook that allows the user to navigate to the different work areas and then back to the beginning.

The result of using this spread sheet is a fully documented process for a solved problem.

Excel Workbook Sample:

Step 1 Example: Problem understanding, and focus must be documented. If it is not written down and clarified the problem solving team may be resolving some problems but they will likely not get to root cause.

Step 1: State the Problem	Return to main page

State the Problem in terms of what you see:

Broken Shaft
Broken bottle
Carton Jams
etc.

1 Problem Statement: Ink Rubs off of package
1 Go see the Problem   Sketch the problem/draw it out
                              Take a very clear picture of the problem
1 Improve the problem statement by stating it as a single Phenomenon. (refer to the book *Stress Free Problem Solving*)
   • simply put it is what you see

Improved Problem Statement: Ink Printing is not bonding to the polyethylene plastic   Go to Step 2:

Phenomenon:
This is the the deviation from what is expected.

The thought process to clarify the problem or phenomenon is to compare the problem situation to the situation when there is no problem. The difference is the phenomenon. It is the problem that must be solved.

Phenomenon Clarification

Problem Statement or phenomenon

A The bottle has fallen and is broken

The effect or phenomenon is different from the expected normal situation.

Expect

Find

The accuracy and clarity of the problem statement is critical.
It is important that the words used reflect accurately the problem that is to be solved to root cause
It is what is seen and nothing else.

Phenomenon "What you see"

The Bottle is broken

Study your current problem statement and see if you can clarify it.   Type in the improved statement below.   Carefully evaluate it

B Broken Bottle

C 2. Define Cause and Effect

Once the problem has been clarified and focused it is time to go on to **Step 2 Cause & Effect**

Step 2 Example:

In step 2 there are five cause and effect categories to examine and define. In the production area equipment, material and sometimes the environment are the primary focus.

In workplaces and offices areas human, systems and sometimes materials are the primary focus.

In laboratories it would depend on the problem focus. If the problem is mechanical then it will mirror the production area, if the problem is work process oriented it would mirror the office area.

The Cause and Effect Worksheet is the document that will be used the most. It will be a summary of the relationship between one cause and the theory of how it creates the unwanted effect. This CE worksheet also has the test plan to prove the cause and effect relationship.

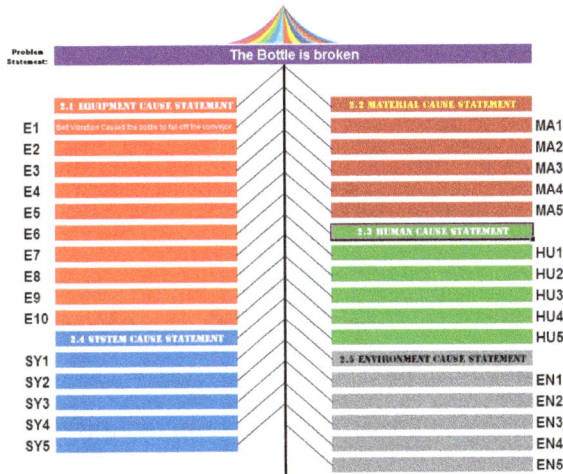

Steps 2.1 through 2.5 provide a one stop location to put in the Cause Statements associated with equipment, material, human, system, and environment.

Step 2.6 Create Cause and Effect proof of Cause worksheets.

This is the PRINT SELECTION AREA

Step 2.6 Provides one location linked to each of the *Proof of Cause* worksheets.

Step 2 Example:

If the relationship is proven to be true the owner of this worksheet will state whether the relationship was true or false. **All** true Cause & Effect relationships go on to the Why-Why analysis.

Example of proof of cause material worksheet five.

All false Cause and Effect relationships are marked if the test shows not meaningful or statistical relationship between cause and effect.

The EquipMent, Material, huMan, SysteM, and EnvironMent links go to all the worksheets.

The cause for each equipment oriented relationship is entered in the equipment cause statement area. There are ten equipment cause statements. If the problem solving team comes up with more than ten equipment causes they should consider.

1. Doing a work point analysis of the problem work point area to get better problem understanding.
2. Revisit the problem statement and clarify the phenomenon.
3. Use the unused human and method worksheets to add their equipment problems.

Stress Free™ Manufacturing Improvement

The Cause and Effect Worksheet is the main problem solving focus. Once this worksheet is filled in, the owner of this worksheet will follow the step by step test plan as written in the proof test plan block.

This worksheet is the backbone of the problem solving process. It facilitates the breakdown of complex, multivariable problems into a set of simple checks to identify the key problem contributors.

This is the time for rigor and discipline. Get it right here and the problem will be solved.

There is a similar worksheet for every cause and effect category.

Step 3 Example:

Each defined cause and effect will need to be field verified to determine if it is true or false. Many problem theories require that certain events or conditions occur. When the condition required to cause the problem can be manually generated, the problem solver should set up the failure condition.

However, this must be done safely.
- **Personal and Equipment safety must be the first concern.**
- **Never create an unsafe or dangerous situation.**
- **Never simulate a problem that risks high financial loss.**

The ability to turn the problem on and off is a signal that the largest contributor to the root cause has been found. Sometimes the cause only creates a portion of the problem. This indicates that it is a multivariable problem with contributing factors that are of similar importance to each other. This is why all true cause and effects are taken into a Why-Why analysis.

Each filled in cause and effect worksheet should be printed out and given to the problem solver that generated the cause and effect theory. The problem solver will then go to the field or area of the problems and follow the test plan. The workbook has each worksheet ready for printing. This is currently one worksheet at a time.

This is normally the point where other personnel must participate to execute the test plan. This participation should have been planned early on and communicated throughout the organization. It is important for all the people affected by the problem to understand what is being done and how they may be asked to contribute to the effort.

Often data must be collected across multiple shifts. Such efforts require some planning, definition of how to collect the data, data collection sheets and some basic training across three to four shifts of personnel.

Do not surprise already hardworking people with the request of extra work! Early engagement will often ensure better problem definition and problem information.

Step 4 Example:

Every true cause becomes the focus for the why-why.

The Why-Why analysis forces the problem solver to think in the reverse order to the Cause & Effect analysis process. This often uncovers additional contributing factors.

Remember: All Why 1's must be field tested before stating a why 2 for each true why 1. This approach maintains a clear logic from true cause to root cause.

Step 4 Example:

Each root cause then must have at least one direct countermeasure. This direct countermeasure must fix the problem. Additional countermeasures may be needed to ensure there is no sliding back into the problem. If the problem has the potential for returning then a periodic check needs to be put into place. This check should be put into the daily inspection standard or the longer term maintenance standard.

71

Here is where many of the items that did not make the cause and effect list show up. They were countermeasures.

Step 5 Example:

The Counter Measures Listed on the Root Cause sheet may need an Execution Plan.

The key is to act to implement the countermeasure in a timely manner.

Do not try to develop a complex execution plan. Sometimes it is just go do.

If multiple problems are being worked on at the same time, the owner of the effort allocation should have an execution plan.

The Counter Measures Listed on the Root Cause sheet may need an Execution Plan. The key is to act to implement the countermeasure in a timely manner.

Counter Measure	Owner	Measure of Success	Mile Stone (MS)							Completion Date
			MS 1	MS 2	MS 3	MS 4	MS 5	MS 6	MS 7	
Corona System Power Back up- Investigate cost	Maria	Report back in two weeks								

Simple example plan format.

Step 6 Example: Stress Free Evaluation

Questions	Evaluation
Equipment	Do / Not Score / OK Mostly / Good / in Clear
1 Equipment operating condition is easy to see.	in Clear
2 Equipment operational and physical settings are easy to set.	in Clear
3 Equipment settings are fixed for each specific product.	in Clear
4 Production variations are easy to prevent.	Not Score
5 Operational centerlines are automatically maintained.	OK Mostly
Material	Do / Not Score / OK Mostly / Good / in Clear
6 Material Quality Characteristics have units and tolerances defined.	Good
7 The supplier is certified and the material Cpk>1.33	Not Score
8 Only logistical damage inspection is required of incoming materials.	Good
9 Damaged incoming materials are not unloaded.	Good
10 In process material issues are rare and easy to recover from.	Good
Human	Do / Not Score / OK Mostly / Good / in Clear
11 Expert problem prevention skills are the norm.	Good
12 Advanced condition management skills are the norm.	OK Mostly
13 Adherence to Standards and following Standard Operating Procedures is	Good
14 Mistake proofing has eliminated mistakes.	OK Mostly
15 Our people are highly motivated, continuous improvement leaders	Good
Systems	Do / Not Score / OK Mostly / Good / in Clear
16 Systems exist and are well documented.	Good
17 Systems ensure Cpk>1.33 and no defects	OK Mostly
18 Systems ensure required production rate is maintained.	Good
19 Systems are explained and trained using the lastest techniques	Good
20 Systems are easy to learn requiring less than five days to get qualified.	OK Mostly
Environment	Do / Not Score / OK Mostly / Good / in Clear
21 Support Organizations accept our solutions and integrate it into their	Good
22 Purchasing adjusts its buying based on the problem solution.	OK Mostly
23 Material specifications are adjusted based on specific problem solutions.	Good
24 Operational conditions are changed as required to maintain improvements	Good
25 Logitics are adjusted to support specific problem resolution.	Good
Leadership	Do / Not Score / OK Mostly / Good / in Clear
26 Leaders review the root cause of all key problems.	in Clear
27 Key problem resolution always gets public recognition.	OK Mostly
28 Leaders schedule root cause validation time.	OK Mostly
29 Leadership goes to the root cause to see and get understanding.	Good
30 Leaders participate in stress free problem evaluation.	in Clear
Savings	Do / Not Score / OK Mostly / Good / in Clear
31 Solving the problem is more important than immediately trying to restart the line.	Not Score
32 Problem Elimination savings are a key concideration	Not Score
33 Profblem cost elimination is evaluated during the budget and elimination goals setting	Not Score
34 Problem cost elimination savings is prioritized and a key improvement concideration	Good
35 Problem Elimination Savings are a key concideration	Good

So, the root cause has been found. A counter measure has been executed. This countermeasure and how long it is kept in place is directly related to how much more work the persons that have daily contact with the problem solution area are required to do.

If the fix made the situation better and the work is less the solution will stay in place for a long time.

If maintaining the conditions for zero loss is difficult and hard to do. The root cause will return in the near future.

Answer the evaluation questions and see the easy to maintain score.

The target should be a minimum of **80 %**.

Visual Stress Free Evaluation

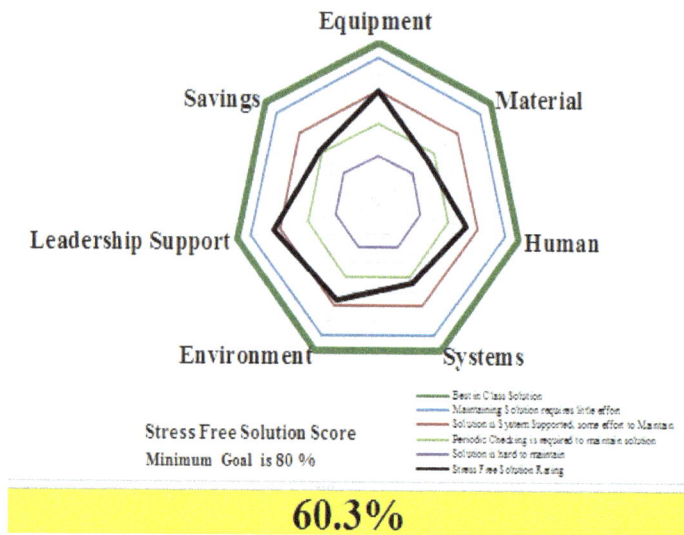

60.3%

Some of the most difficult root cause problems that I have solved and felt so good about ended up with very low easy to maintain scores. The low scores are an indication that additional improvement is needed, and that the problem area will need to be monitored.

Step 7 Example: Document, Standardize, Train

If this step is not done well, the ability to maintain the root cause solution is put at risk. Training is most often addressed but the upgrade of standards and the system documentation is often overlooked. Missing any of these areas puts the solution in jeopardy.

1. Update all Documentation:

All business, control documents, drawings etc. must be upgraded to reflect the solution. Not doing so jeopardizes the final outcome.

Examples: Engineering Equipment Drawings

- Maintenance Procedures
- Operating Procedures
- Clean, Inspect, Lubrication procedures
- Safety guidelines and procedures
- Set up and Start up procedures
- Spare parts purchasing specifications and amounts

3. Establish and Maintain the use of Standards:

   Document the new or improved standard(s) to ensures the problem will never occur again.

   Examples:
   - Quality Standard
   - Safety Standard

4. Establish a Qualification and Periodic Requalification System

5. Train all those who must maintain the countermeasure.
   - All Operational Teams
   - All Maintenance Teams
   - All Engineering resources
   - All Leadership

Ron Mueller

Ron Mueller P.E.

- Integrated Work Systems (IWS) materials author
- Coach to dozens of Manufacturing Directors across the world.
- Certified TPM Coach.
- Tested and proven to enable true breakthrough improvement of Supply Chains.

A proven leader of smart systems implementation across supply, manufacturing, and distribution to drive out cost, inefficiencies and to establish synchronized Supply Chains. He utilized the best thinking of Japan's TPM leaders and crafted the necessary related pillars and systems that work in Consumer Products Manufacturing. The results delivered include reduction of Raw and Finished Product Inventories by 40%. Delivered over $100 million is loss reduction through focused systems Workshops across dozens of sites. Developed P&G IWS program materials for external sale. Winner of P&G's Diamond Award for Contribution to Product Supply.

Core Competencies include:

✓ Coaching Manufacturing Leadership,

✓ Implementation of Integrated Work Systems,

✓ Statistical Replenishment design and implementation,

✓ Supply Chain Synchronization: author of 3 books in the Stress Free^{TM} series that aid Business and Supply Chain leaders to develop and improve their organization's performance.

Ron Mueller

Gordon Miller P.E.

- Manufacturing Performance Program
- Development and Delivery Expert.
- Application of Intelligent Manufacturing technology against biggest business challenges with proven business results.

A record as a collaborative and leading-edge thinker, developing programs to deliver cost, productivity and growth enabling manufacturing technology systems deployed via smart standards and empowered teams. As an early developer of PR/OEE measures and improvement programs, has experience with unlocking organization capability for improvement with smart strategies. Led program that developed initial P&G Manufacturing Execution System, leveraged globally across multiple GBUs. Influenced Beauty and Household Care manufacturing systems changes that enabled and leveraged global standardization for rapid footprint growth. Experience that enabled 50% reduction in OEE losses. Experience as a leader of corporate STEM talent strategy can assess and devise approaches to ensure Talent needs for the challenging future are met.

Core Competencies include:

- ✓ Global Productivity Program Design and Management,
- ✓ Advanced Manufacturing Technology Innovation and Strategy Development,
- ✓ Development of Highly Effective Global Teams,
- ✓ Vendor development and management, Organization Capability Development,
- ✓ Talent Strategy

Design of the Stress Free Solutions Suite

Stress Free Manufacturing Solutions is available in two parts:

Stress Free Manufacturing Solutions – the book

- Presents the theory
- It comes as an e-book or
- Seven by Ten Paperback copy or

    The e-book is free with the purchase of the hard copy

Stress Free Manufacturing Solutions - Excel Workbook

- Provides an "easy to document" environment
- This provides the user a way to optimize their resources and time.

Other Books by Ron Mueller

Stress Free™ Supply Chain Solutions

Stress Free™ Manufacturing Solution

Stress Free™ Work Process Solutions

Stress Free™ Changeover Solutions

Stress Free™ Daily Management Solutions

Around the World Publishing LLC

4914 Cooper Road Suite 144

Cincinnati, Ohio 45242-9998

www.ingramcontent.com/pod-product-compliance
Lightning Source LLC
Chambersburg PA
CBHW081747200326
41597CB00024B/4418